MW01641014

Let's Explore Night and Day

by Laura Hamilton Waxman

Bumba Books™

LERNER PUBLICATIONS ◆ MINNEAPOLIS

Note to Educators

Throughout this book, you'll find critical-thinking questions. These can be used to engage young readers in thinking critically about the topic and in using the text and photos to do so.

For Greta, Signe, and Leif, who bring joy day after day

Copyright © 2022 by Lerner Publishing Group, Inc.

All rights reserved. International copyright secured. No part of this book may be reproduced, stored in a retrieval system, or transmitted in any form or by any means—electronic, mechanical, photocopying, recording, or otherwise—without the prior written permission of Lerner Publishing Group, Inc., except for the inclusion of brief quotations in an acknowledged review.

Lerner Publications Company
An imprint of Lerner Publishing Group, Inc.
241 First Avenue North
Minneapolis, MN 55401 USA

For reading levels and more information, look up this title at www.lernerbooks.com.

Main body text set in Helvetica Textbook Com Roman.
Typeface provided by Linotype AG.

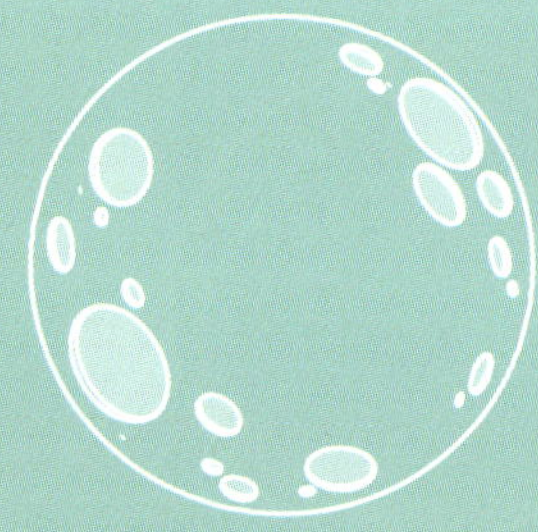

Library of Congress Cataloging-in-Publication Data

Names: Waxman, Laura Hamilton, author.
Title: Let's explore night and day / Laura Hamilton Waxman.
Description: Minneapolis : Lerner Publications, [2022] | Series: Bumba books - Let's explore nature's cycles | Audience: Ages 4–7 | Audience: Grades K–1 | Summary: "Learn the names for the different parts of the day and night cycle, beginning with dawn and ending with night"— Provided by publisher.
Identifiers: LCCN 2020004914 (print) | LCCN 2020004915 (ebook) | ISBN 9781728404028 (library binding) | ISBN 9781728417714 (ebook)
Subjects: LCSH: Earth (Planet)—Rotation—Juvenile literature. | Night—Juvenile literature. | Day—Juvenile literature.
Classification: LCC QB633 .W38 2022 (print) | LCC QB633 (ebook) | DDC 525/.35—dc23

LC record available at https://lccn.loc.gov/2020004914
LC ebook record available at https://lccn.loc.gov/2020004915

Manufactured in the United States of America
1-48460-48974-10/12/2020

Table of Contents

A New Day

A new day begins each morning. Darkness comes each night. It's a cycle that repeats again and again.

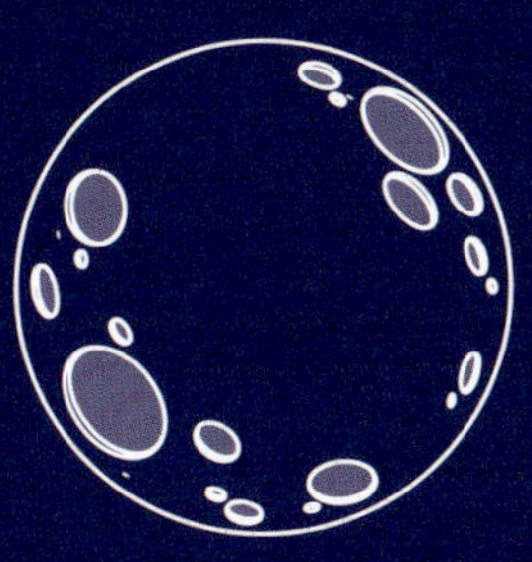

It's dawn.

The morning sky

begins to brighten.

The sun is still below

the horizon.

The sun comes out.

Bright orange and yellow clouds sweep across the sky.

It's a beautiful sunrise.

At midday, the sun is high in the blue sky.

It warms the air around us.

Why do you think day is warmer than night?

Evening is here, and so is the sunset. The sun sinks low in the sky. The air cools down.

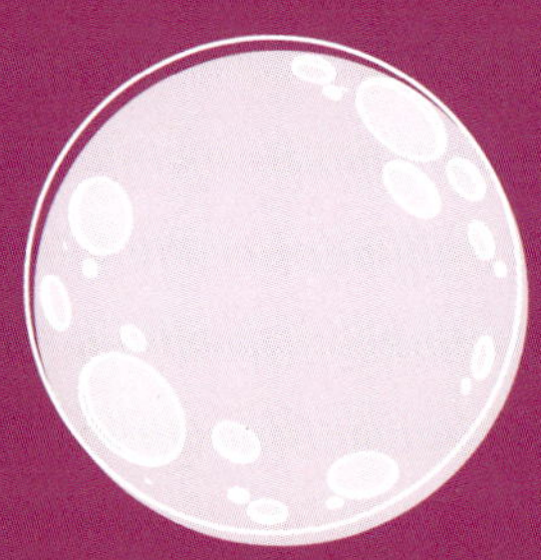

It is dusk.

The sun has gone below the horizon.

The sun's light and warmth are fading.

Where do you think the sun goes at dusk?

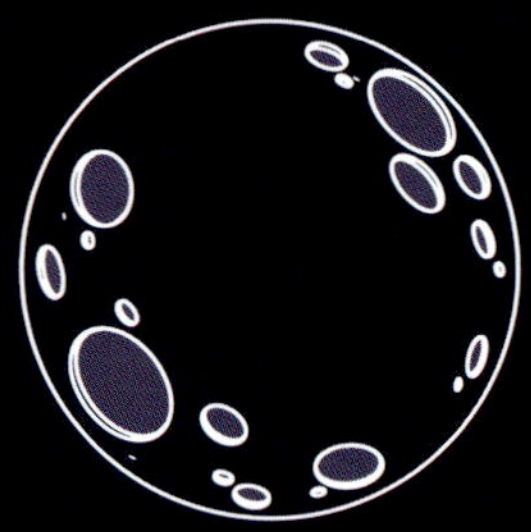

Night has come.

Stars shine in the dark sky.

The moon is bright.

The moon seems to change shape from night to night. Some nights it looks bigger. Other nights it looks smaller.

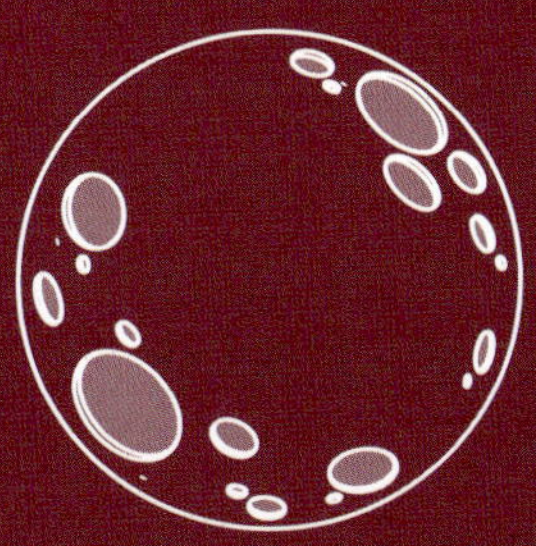

It's a new day.

The morning sun lights up the sky.

The cycle of day and night begins again.

The Cycle of Day and Night

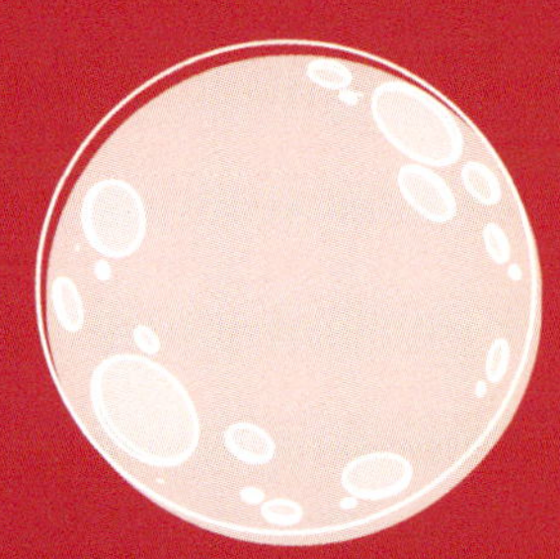

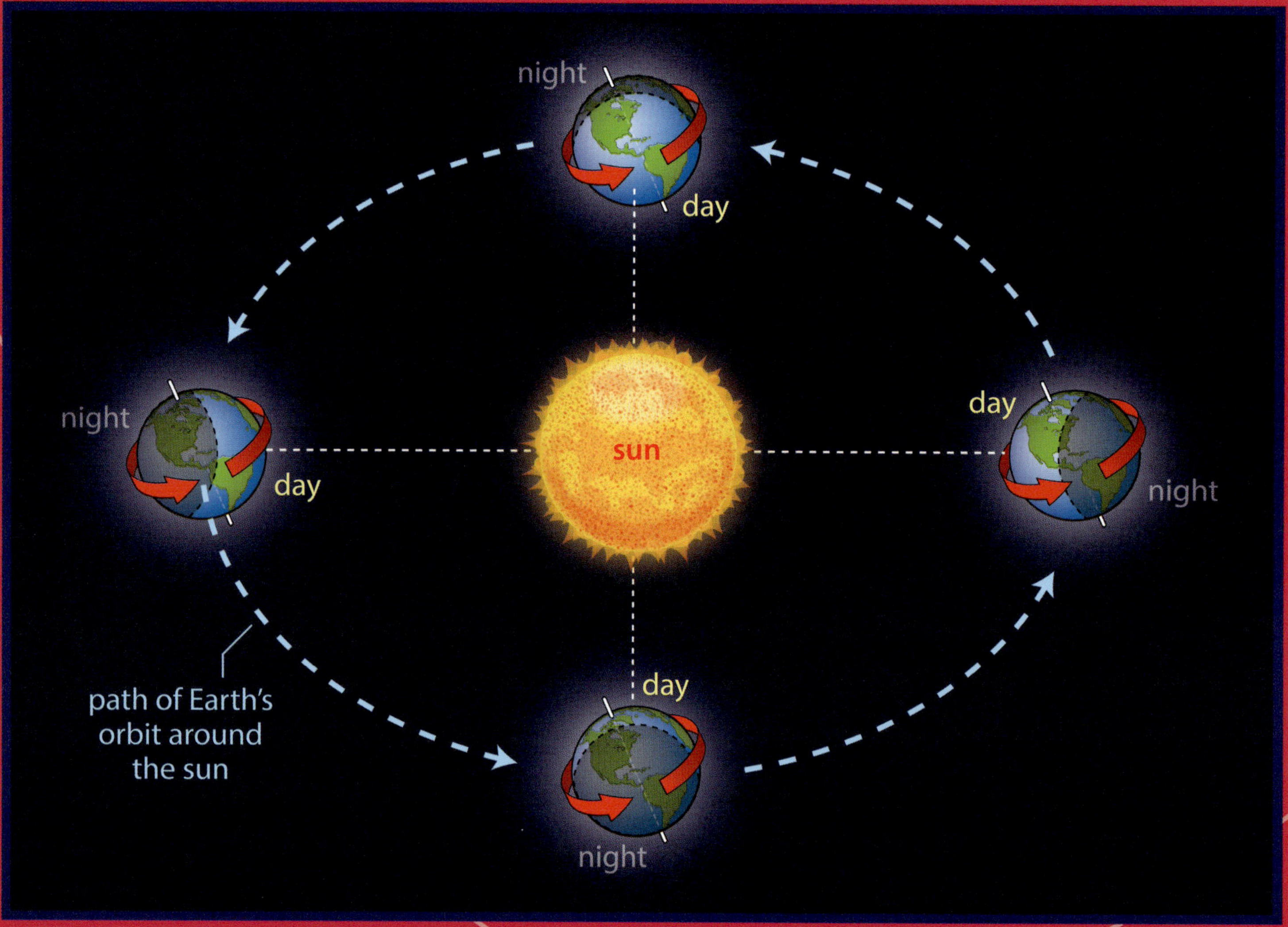

night
day
night
day
sun
day
night
day
night
path of Earth's orbit around the sun

Picture Glossary

cycle

something that repeats again and again

dawn

the end of the night when the sky begins to brighten

dusk

the end of the day when the sun is too low in the sky to see

horizon

the line where the sky seems to meet the land or sea

Learn More

Kim, Mi-hye. *Day and Night*. Minneapolis: Big & Small, 2016.

Rustad, Martha E. H. *Does the Sun Sleep? Noticing Sun, Moon, and Star Patterns*. Minneapolis: Milbrook Press, 2016.

Waxman, Laura Hamilton. *Let's Explore Phases of the Moon*. Minneapolis: Lerner Publications, 2022.

Index

Photo Credits

Image credits: saravutpics/Shutterstock.com, pp. 5, 23 (top left); Rebecca Nelson/Getty Images, pp. 6, 23 (top right); Jacob_09/Shutterstock.com, p. 9; Hillary Kladke/Getty Images, p. 10; Iriskarightnow/Shutterstock .com, pp. 13, 23 (bottom right); Oilgraph/Shutterstock.com, pp. 14, 23 (bottom left); Zacarias Pereira da Mata/ Shutterstock.com, p. 17; IgorZh/Shutterstock.com, p. 18; \ Kimberly Boyles/Shutterstock.com, p. 21; Laura Westlund/Independent Picture Service, p. 22.

Cover: Suppakij1017/Shutterstock.com; JIMKOJI/Shutterstock.com.